HUMAN SENSES

Izzi Howell

Published in paperback in Great Britain in 2020 by Wayland
Copyright © Hodder and Stoughton, 2017

All rights reserved
ISBN: 978 1 5263 0683 8
10 9 8 7 6 5 4 3 2 1

Wayland
An imprint of Hachette Children's Group
Part of Hodder & Stoughton
Carmelite House
50 Victoria Embankment
London EC4Y 0DZ

An Hachette UK Company
www.hachette.co.uk
www.hachettechildrens.co.uk

A catalogue for this title is available from
the British Library
Printed and bound in China

Produced for Wayland by
White-Thomson Publishing Ltd
www.wtpub.co.uk

Editor: Izzi Howell
Design: Clare Nicholas

Picture and illustration credits:
Alamy: Paulo Oliveira 15t; Getty: SerrNovik 4, tepic 5t, Tom Brakefield 5b; Shutterstock: Rosa Jay *cover tr*, Anan Kaewkhammul *cover bl*, Sergei Kolesnikov *cover br*, Sergei Kolesnikov *title page* and 18tl, jehsomwang 6 and 10, Paul Looyen 7t, Daxiao Productions 7bl, Eric Isselee 7br, 9tl and 19tr, Smit 8t, Polarpx 8b and 9b, Ekaterina V. Borisova 9tr, Potapov Alexander, schankz, Stephanie Zieber, Nikolai Tsvetkov, Eric Isselee, Littlekidmoment 11t-b, Kalmatsuy 12, QiuJu Song 13t, ilikestudio 13b, Jiri Hera 14tl and 14tr, AlenKadr 14cl, aperturesound 14cr, Lukas Gojda 14b, Ulrich Mueller 15b, TinnaPong 16t, Kjersti Joergensen 16b, Boule 17t, photka 17c, Konstantin Gushcha 17bl, Mariolla1 17br, sunsetman 18tr, KOO 18b, Andrey Pavlov 19tl, Christopher Wood 19b, worldswildlifewonders 20t, Tomas Kotouc 20b, tranac 21.
All design elements from Shutterstock.

Should there be any inadvertent omission, please apply to the publisher for rectification.

The author, Izzi Howell, is a writer and editor specialising in children's educational publishing.

Contents

What are senses?	4
Sight	6
Looking around	8
Hearing	10
Taste	12
Different tastes	14
Touch	16
Smell	18
Special senses	20
Human and animal senses	22
Index	24

What are senses?

Humans and animals use their senses to find out information about the world around them.

sight

hearing

smell

taste

touch

Senses keep humans and animals safe. We know when danger is coming.

Hurry up!

a car

a dangerous predator

Which senses would these people use to know that a car is nearby?

predator: an animal that kills and eats other animals

Sight

Humans and animals use their eyes to see.

Light from an object goes into the eye through the pupil.

The light makes an upside-down image at the back of the eye.

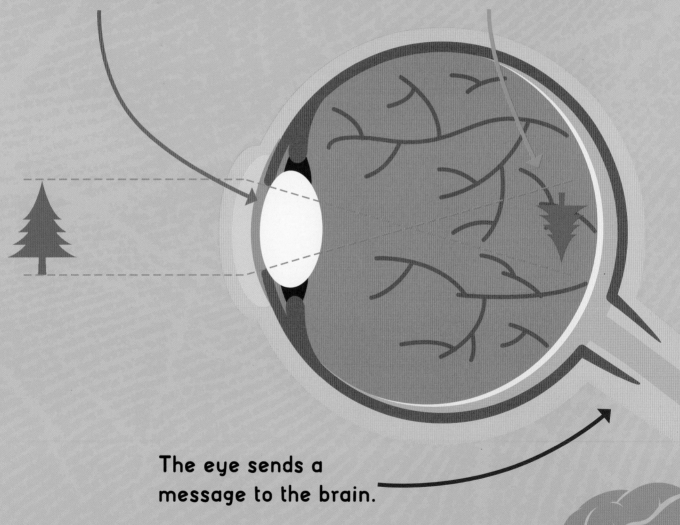

The eye sends a message to the brain.

The brain turns the image the right way up and understands what you are seeing.

pupil: the black, round hole in the centre of the eye

Animal eyes come in different shapes and sizes. Some animals have more than two eyes.

a spider

eight eyes of different sizes

two small, oval eyes

two big, round eyes

an eagle owl

a human

Looking around

Predators usually have eyes at the front of their head. They can see what is straight in front of them.

a tiger

Roar!

Seeing only in front helps predators focus on finding prey.

prey: an animal that is killed and eaten by another animal

Prey animals usually have eyes on the sides of their head. They can see in front and to the sides.

a deer

a robin

Seeing a bigger area helps prey spot predators faster.

Hearing

Animals and humans hear sounds with their ears.

Sound waves go into the ear.

The waves hit the eardrum and make it vibrate.

The vibrations travel through the middle and inner ear.

The inner ear sends a message to the brain.

The brain understands what you are hearing.

vibrate: to quickly move backwards and forwards

a dolphin

better

a mouse

Some animals can hear much better than humans. They can hear faraway sounds, and high and low sounds that we can't hear.

a cat

a dog

HEARING

Which animal can hear more sounds, a cat or a sheep?

a sheep

a human

good

Taste

Humans and some animals sense taste with their tongue.

Taste buds on our tongue recognise different tastes (see page 14).

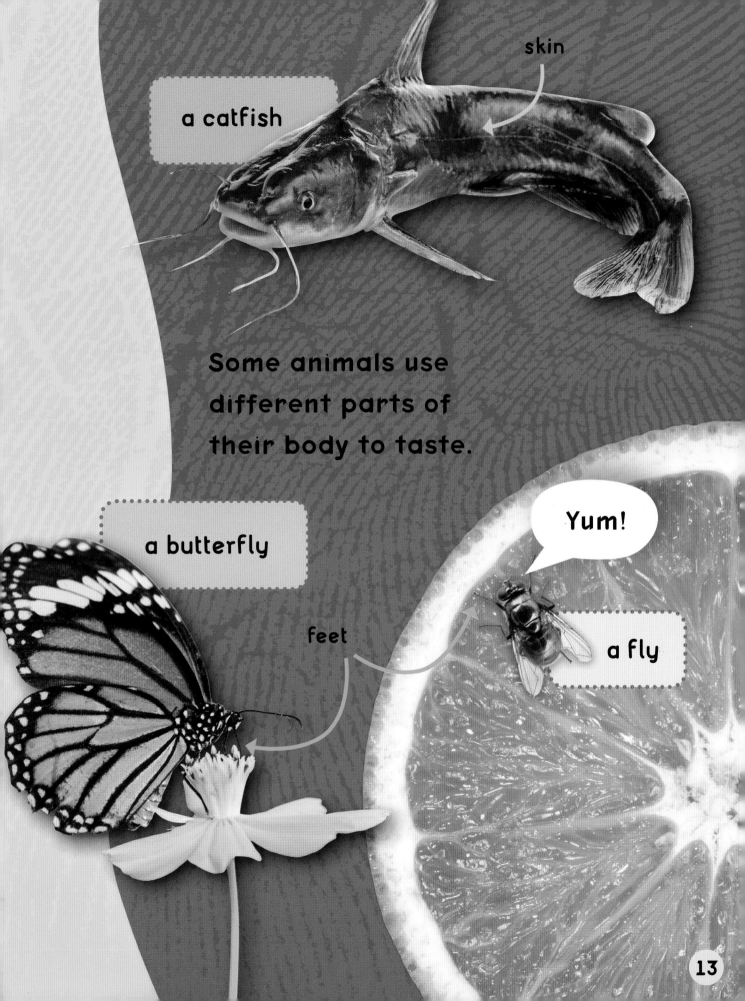

a catfish

skin

Some animals use different parts of their body to taste.

a butterfly

feet

Yum!

a fly

Different tastes

Humans taste five different flavours with their tongue.

sweets

sweet

salt

salty

Most foods have a combination of these five flavours. Which flavours can you taste in a fruit salad?

lemons

sour

soy sauce

umami (rich and meaty)

dark chocolate

bitter

Octopuses taste food with the suction cups on their legs. They don't eat food if it doesn't taste nice.

suction cup

Delicious!

Cows spit out plants that don't taste nice to them. Plants that taste bad are often poisonous.

poisonous: something that can make you sick if you eat or drink it

Touch

Humans and some animals are covered in skin. We can sense when something touches our skin.

whiskers

Whiskers are also sensitive. They help some animals to feel what is around them.

When we touch an object, we can feel if it is...

...cold,

...rough,

What other textures can you feel by touching something?

...slimy,

...furry.

Smell

Humans and some animals smell with their noses. Some animals smell using other parts of their bodies.

antennae

nose

a crab

a human

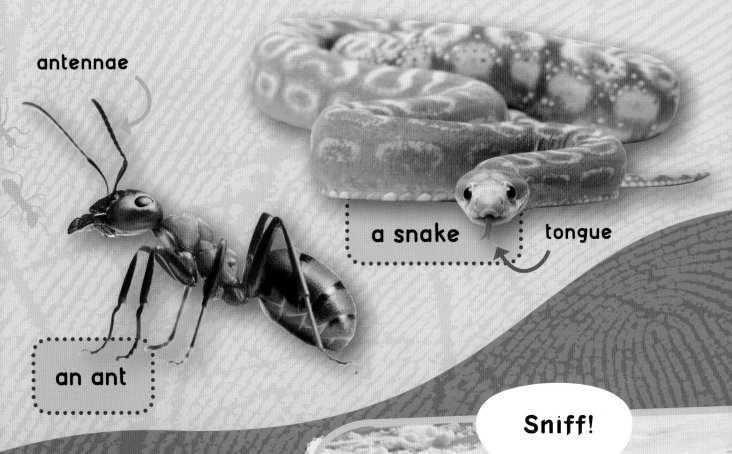

antennae

a snake   tongue

an ant

Animals use smell to find food. They can also smell if other animals are nearby.

Sniff!

A mole uses smell to catch earthworms.

This polar bear is using smell to hunt seals.

Special senses

A few animals have extra senses that humans don't have.

Some animals find food underwater by sensing electricity in the bodies of prey animals.

a platypus

a lemon shark

Vampire bats and some snakes can sense the heat of other animals. This helps them to find prey to eat.

> Which senses do humans use to know if something is hot?

This is how a snake would sense a kangaroo.

Human and animal senses

Most mammals and birds have the same five senses as humans.

Mammals

cat
dog
cow
tiger
deer
dolphin
mouse
sheep
human
mole
polar bear
platypus
vampire bat

Birds

eagle
owl
robin

eyes – see
ears – hear
tongue – taste
nose – smell
skin – touch

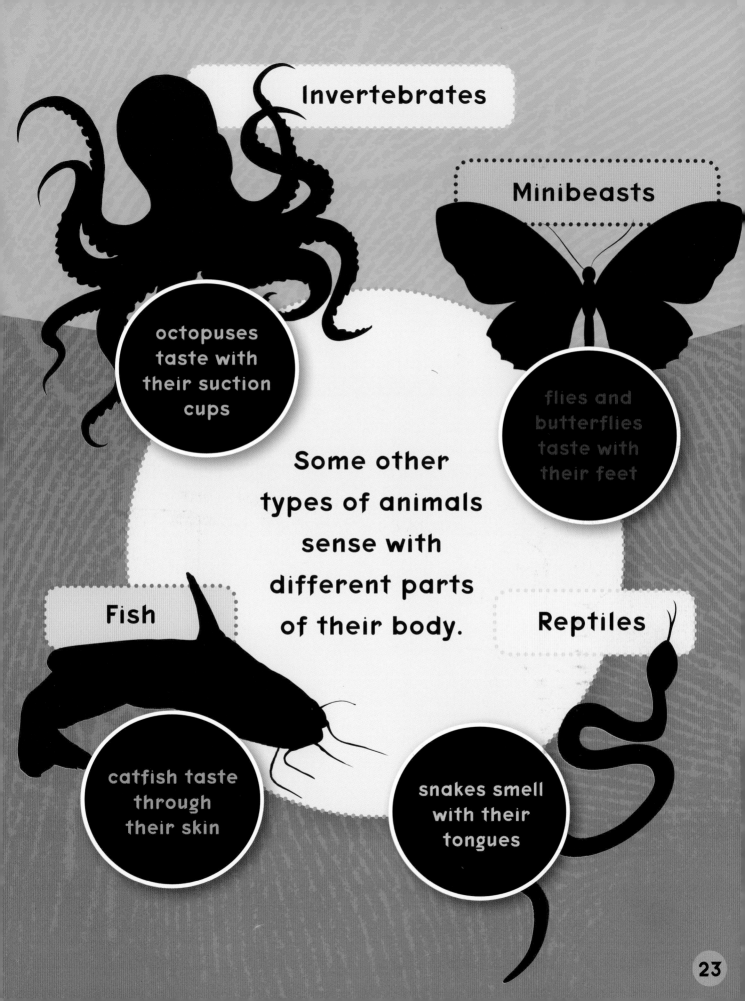

Index

antennae 18, 19

birds 7, 9, 23
brains 6, 10

ears 10,
eyes 6, 7, 8, 9

feet 13
fish 13, 20, 23
flavours 14–15

hearing 4, 10–11
humans 4, 5, 6, 7, 10, 11, 12, 14, 16, 18

insects 7, 13, 19
invertebrates 15, 18, 23

mammals 4, 5, 8, 9, 11, 12, 14, 15, 16, 18, 19, 20, 21, 23
minibeasts 23

noses 18

predators 5, 8, 9

prey 8, 9, 21

reptiles 19, 21, 23

sight 4, 6–7, 8–9
skin 13, 16
smell 4, 18–19
special senses 20–21

taste 4, 12–13, 14–15
tongues 12, 19
touch 4, 16–17

whiskers 16

Answers

p5 — Sight, hearing
p11 — Cat
p14 — Sweet and sour
p21 — Touch, sight, hearing